HUDSON 70

In Words and Pictures

First circumnavigation of the Americas:

An 11 month Canadian voyage of discovery

that traversed four oceans

A collection of pictures and perspectives about the expedition compiled by two of its participants

Dr. Charles T. Schafer Pgeo(NS)
Emeritus Research Scientist
Geological Survey of Canada – Atlantic
Bedford Institute of Oceanography (BIO)
Dartmouth, Nova Scotia B2Y 4A2

Mr. Roger W. Smith,
President
Big Muddy Exploration Ltd.
Suite 555, 444 5th Ave SW
Calgary, Alberta T2P 2T8

"We should celebrate Hudson 70 … for daring to do great things in the ocean and thus to inspire the scientists on board to think great thoughts. It is wonderful that Hudson is still afloat at the advanced age of 48, and still doing valuable work in the waters around Canada."

— Professor Peter Wadhams
Cambridge University, UK, September, 2009.
(Professor Wadhams was a Hudson 70 participant)

Dear Reader,

Captions for the photos presented in this book can be found after the last chapter. Try guessing what each photo represents. Many are easy to identify but others are not.

Questions and comments about this book and the material presented can be sent to **charlestschafer@gmail.com**

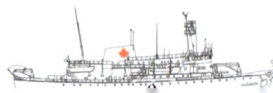

Acknowledgements

We are indebted to a number of colleagues and contacts that supplied critical information about particular parts of the *Hudson 70* expedition. Mr. Jim Shearer (Ottawa), Dr. Vaughn Barry (Pacific Geoscience Centre) and Ms. Pam Wilkins (library, Institute of Ocean Sciences, Sidney, BC.) provided needed information and photographs for the chapter that describes the support roles played by the *Richardson* and the *Endeavour*. Dr. Iver Duedall (Florida Institute of Technology), Mr. Kelly Bentham (BIO - Dept. of Fisheries and Oceans) and Mr. David Frobel (Natural Resources Canada—Geological Survey of Canada-Atlantic) resurrected some excellent photographs of everyday laboratory and other shipboard activities that were carried out during the expedition.

We would also like to offer our special thanks to Mr. Kevin MacIssac and his fellow biologist colleagues at BIO (Dept. of Fisheries and Oceans) for their assistance in identifying some of the species that were found in our personal photograph collections. Last but not least, our sincere appreciation to our draft manuscript reviewers Mr. Randy Currie, Mr. Glen Faucher, Mr. Tom Mainville, Mr. Patrick Potter and Dr. B. Loncarevic. Ms. Deborah Perry, a Natural Resources Canada graphic designer volunteered to rework our final draft giving it a style that has made it much more appealing for a general public readership.

Dedication

This book is dedicated to the scientists and crew of the *Hudson 70* expedition and to the many other 20th century Canadian scientists, departmental managers and federal government politicians that either directly or indirectly helped to make this unique Canadian voyage of discovery a reality.

— INTRODUCTION —

HUDSON 70:

THE IDEA AND THE PLANNING

On 19 November 1969, the *CSS Hudson* steamed out of Halifax Harbour on what was to become the first circumnavigation of the Americas by a research ship. The *Hudson 70* expedition was divided into 9 legs and 11 scientific program phases with ports of call on both the Atlantic and Pacific sides of South America, the island of Tahiti, Vancouver, and at places in the Arctic such as Tuktoyaktuk and Resolute where crews were exchanged by helicopter and launches. The voyage was among the last of a series of big multi-disciplinary global oceanographic expeditions that occurred during the 19th and 20th centuries. The final legs of the expedition carried vessel and crew through the Northwest Passage and Baffin Bay along a track where no research vessels had gone before. However, there were many other equally spectacular and scientifically interesting places encountered during the course of *Hudson's* 104,000 kilometer 11 month journey to far-flung areas of the four oceans that surround North and South America. The expedition took place at a time when assessment of the marine environment on a global scale was becoming of increasing importance to governments and environmental organizations alike. During the various legs of the voyage, *Hudson* carried a total of 122 scientific and technical staff from several countries and returned home with an impressive collection of samples and observations that would become the subject of scientific publications for decades to come. Such an ambitious undertaking had not been witnessed in Canada since 1954 when the *HMCS Labrador*, under the command of O.C.S. Robertson, circumnavigated North America to become the first icebreaker to traverse the Northwest Passage.

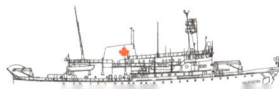

Hudson is one of Canada's oldest operational government ocean research vessels. Built in New Brunswick in 1963 by St. John Shipbuilding and Dry Dock Ltd., the 90.4 m long, 4700 ton ice-reinforced vessel has a cruising range of about 23,000 km. To fulfill her scientific and marine survey roles, there are three laboratories, a helicopter deck, three lifting cranes, and three hydrographic winches that can quickly lower instruments and sampling gear to depths of more than three km.

The *Hudson 70* expedition's track around the Americas is reminiscent of earlier lengthy voyages of discovery that have taken place over the past several centuries. During the first decades of the 17th century, the 12 m long 20 ton "fly-boat" *Discovery* owned by the British East India Company was used initially to bring colonists to the new colony of Virginia in 1607. Several years later, *Discovery* participated in six expeditions in search of the Northwest Passage. Another voyage of note was made by Captain James Cook on the 368 ton 32 m long British Royal Navy research vessel *HM Bark Endeavour*. His almost three years long voyage of discovery brought him to the coasts of Australia and new Zealand between 1769 and 1771. Captain Cook is credited with completing the first map of the coasts of Newfoundland and the entrance of the St. Lawrence River. The river survey was carried out during the siege of Quebec and is reported to have allowed General Wolf to make his famous surprise attack on the Plains of Abraham.

The 235 ton 28 m long *HMS Beagle* was launched in 1820 and especially commissioned

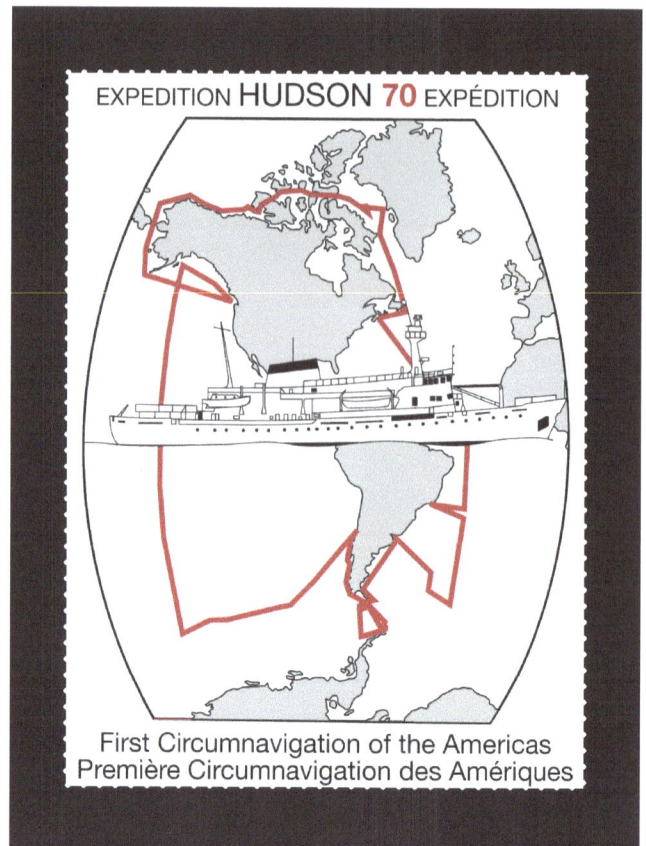

EXPEDITION **HUDSON 70** EXPÉDITION

First Circumnavigation of the Americas
Première Circumnavigation des Amériques

for the British Navy's survey program. For 17 years, Beagle surveyed much of the world including the coasts of South America (1826 to 1835) and many of the small islands scattered throughout the Pacific (1835 to 1840). Charles Darwin joined the ship in 1831 as a naturalist and, during various Beagle voyages, developed his theories on the origin and evolution of species. One of the most famous of the late 19th century voyages of discovery was made by the 69 m British Navy vessel *HMS Challenger*. She circumnavigated the ice-free oceans of the world between 1872 and 1876 making systematic observations of ocean water characteristics, sounding the ocean's depths and collecting hundreds of new marine species.

Challenger visited Halifax for a week or so in May 1873 before resuming her journey to the coasts of Africa and South America. At that time, voyages of discovery were still relatively general in nature. For example, the *Challenger's* mission was to investigate the distribution of animals in the deep sea and to observe how the oceans circulate. It was a classic example of a multidisciplinary expedition that produced a large amount of new information on the temperature and salinity of seawater, and on the distribution of deep ocean sediments and species. Publication of *Challenger's* findings took about 20 years and filled 50 volumes totaling about 30,000 pages.

Subsequent late 19th and early 20th century expeditions tended towards more of a targeted approach. For example, voyages undertaken by the 36 m long 420 ton *Fram* to the Arctic in 1893, and others made by the 52 m long 736 ton *RRS Discovery* to the Antarctic (1901-1904) and by the *USNS Eltanin* in 1962, were tasked with relatively focused geographic objectives aimed at increasing the knowledge of the biology and ecology of the seas surrounding the Arctic and Antarctica. As such, the *Hudson 70* broad-brush mission was in keeping with the tradition of earlier voyages of discovery. In retrospect, the expedition was exactly what was needed at the time to establish Canada's credentials as a nation with a strong interest in "blue water" ocean research.

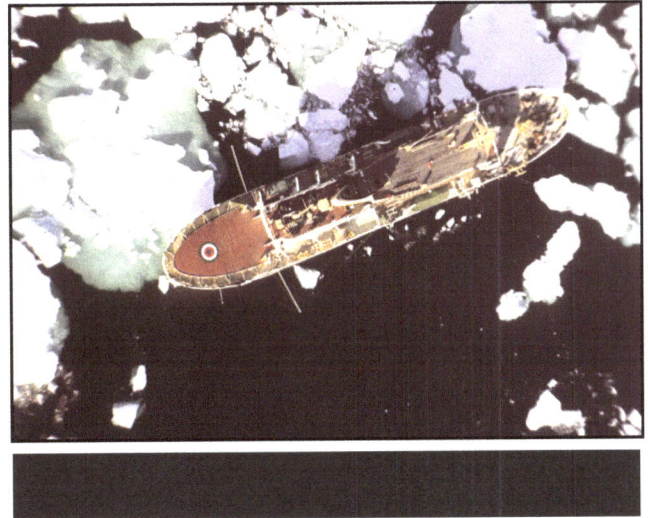

Captain David Butler   Dr. C. R. Mann   Dr. W. L. Ford   Dr. C. D. Maunsell

Big ideas often develop from a few smaller ones and the *Hudson 70* concept appears to have followed that model.

Hudson was commissioned in 1964. By 1967, the ship had been used as a platform for marine geology and geophysical investigations along Canada's eastern seaboard. It had been suggested that she should also be deployed to the west coast continental shelf and the Beaufort Sea to explore these areas with the sophisticated geophysical instruments that had been installed for her east coast surveys. During the 1960's, marine scientists were busy developing equipment that could offer new insights into the properties of ocean currents, seawater chemistry and the biological life of the oceans. It was considered important to evaluate these prototypes and new techniques on an ocean-wide scale to gain new understandings about the complete ocean system. These ideas eventually formed the basic framework of what would become the *Hudson 70* expedition. By 1968, BIO scientists were convinced that the best way to accomplish both ocean-scale and Canadian-focused regional objectives would be to send the ship on a precedent-setting voyage around North and South America. In this way, Canadian scientific objectives could be achieved and, in the process, a major contribution to knowledge of the world's oceans and an important element of Canada's contribution to the International Decade of Ocean Exploration (1970-1980) would be realized.

By August 1968, a detailed scientific program was ready. It was subsequently approved in the following November by the Hon. J.J. Greene, Minister of the Department of Energy Mines and Resources. A coordinating committee was set up on November 29th. It consisted of nine members and was chaired by Dr. C.R. Mann who was one of the key supporters for the expedition. The final cost of the expedition was about 1.6 million dollars and involved a total of 45 person years.

— Collecting This and That —

❧

The Wish Lists of Scientists Were Ambitious

Three key activities of most marine research expeditions are sampling, direct measurements of physical parameters and remote sensing. All three featured prominently in the 11 phases of the *Hudson 70* scientific program. Oceanographers that participated in one or more of the expedition's 9 legs had questions about the physical and chemical characteristics of sea water over a wide range of water depths. This information was needed to define the occurrence and geometry of both large water masses and major deep sea ocean currents. Two important ocean water characteristics needed for this kind of study are temperature and salinity. The water samples needed for both salinity and for other chemical analyses were collected using various types of bottle samplers.

The standard water sampler at the time of the voyage was the Knudsen bottle. It was made from a non-corrosive metal alloy and featured a tube-shaped body that collected the water sample. Several smaller external perforated tubes attached to the body held glass reversing thermometers. The bottle looked like a short metal 8 cm diameter pipe with "flip top"-type caps and cable clamps at each end. A bottle was clamped to a thin wire cable at various intervals. The cable was lowered until its anchor weight reached the appropriate depth. The bottle's spring-loaded end caps were activated using trigger weights that ran down the cable between each bottle. Samples were used to construct a profile of sea water properties at a particular location and at the depths that the bottles were tripped.

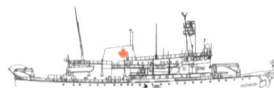

The end cap closing action of the bottle is accompanied by two other events. One of these is the release of a second trigger weight attached to the bottom of the first sampler that slides down the wire to close the next bottle. The second weight-triggered event is the release of a pin that allows the bottle to flip 180 degrees so that it is hanging upside down by its lower attachment clamp. The sequence is repeated until the deepest bottle has been closed and flipped. Flipping the bottle locks the position of the mercury column of the glass reversing thermometers giving the temperature of the water at the depth where the sample was collected. Larger bottles (e.g., the Niskin bottle) are made from PVC and other non-metallic components. They were used during the expedition to collect water samples for chemical analysis of trace metals, organic pollutants and of gasses dissolved in sea water.

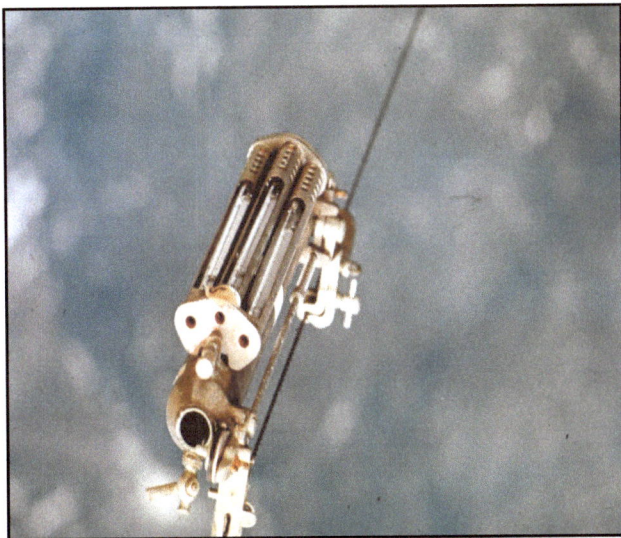

The tool kit of *Hudson 70* oceanographers included several other instruments that contributed to obtaining a more comprehensive picture of the water masses under investigation and their dynamics in poorly known parts of the ocean's basins. One of these is the expendable bathy-thermograph or XBT. The XBT launcher used during the expedition looks like a length of 5 cm diameter pipe. After the XBT probe is released from the launcher, it falls through the water. Two very fine wires that are connected to recording instruments on the ship uncoil from their storage spools within the probe during its decent. They conduct signals back to the ship giving a virtually continuous record of temperature change as a function of depth. These data were used to fill in the gaps between temperatures recorded at specific depths by the reversing thermometers mounted on the body of Nansen bottles.

Measurement of subsurface and deep ocean currents during the expedition was done using current meter moorings. Each mooring consisted of a large orange-colored sub-surface float linked to a long wire cable that was used to mount a number of recording current meters that were attached at various intervals along the cable. The mooring was streamed behind the ship and then its anchor weight (a large railway car wheel weighing several hundred kilograms) was lowered over the side of the ship and released. The mooring float's attached parachute insured its slow decent to the seafloor. The railway car wheel was linked to the mooring by an acoustic release unit that was triggered from the ship when the time had come to recover it. The float lifted the mooring back to the surface for retrieval. Because moorings can drift away from their anchored location during accent, floats were typically fitted with flashing strobe lights and a radio beacon to assist in relocating them for recovery once they had surfaced.

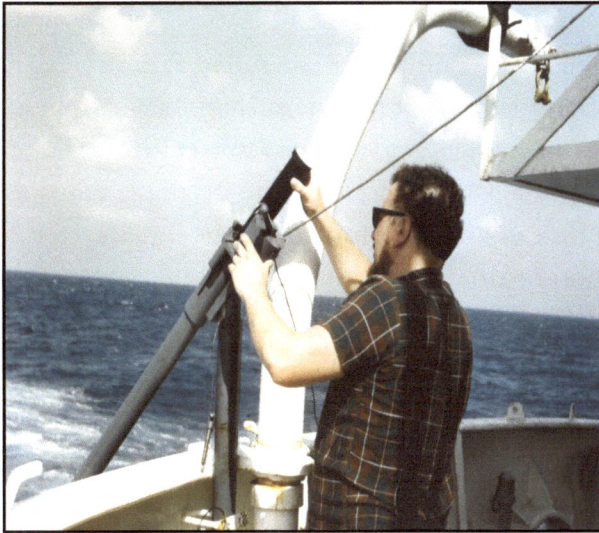

Initial results of the *Hudson 70* oceanographic program offered a snapshot of the two-dimensional distribution of water masses along the ship's track through some remote parts of the Atlantic and Pacific oceans as well as new information on the concentration of carbon dioxide, dissolved oxygen and nutrients. These results would be compared with earlier and future measurements to show how the ocean's characteristics have varied over time and how they might be influenced by climate change.

The deployment of a number of current meter moorings across the Drake Passage off the southern tip of South America gave the first evidence about the rates and directions of water flows between the Atlantic and Pacific oceans via this southern pathway.

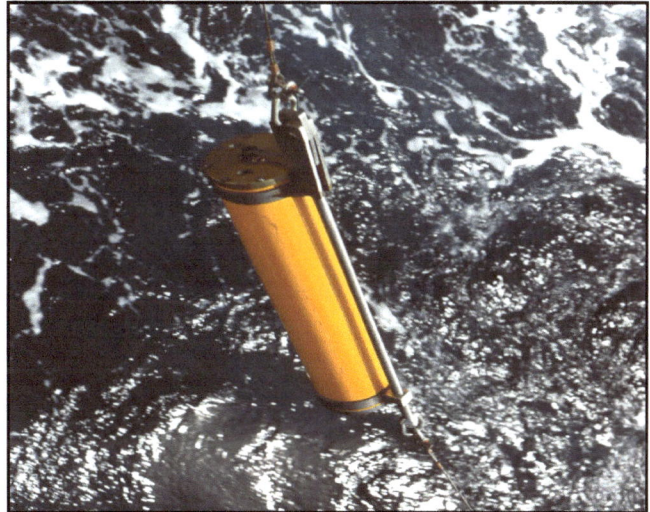

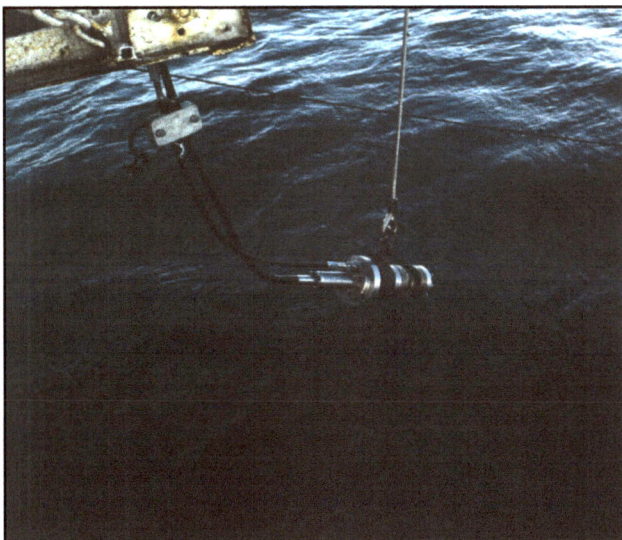

Hudson 70 marine geologists wanted to know more about the composition and distribution of deep ocean seafloor sediments while their marine geophysicist counterparts had posed a number of questions about the thickness of deep sea sediment deposits in relation to their underlying bedrock foundations, especially in some poorly known areas of Canada's Pacific and Arctic offshore.

Geologist's equipment included a number of different surficial sediment samplers, sediment coring devices and underwater cameras. Their geophysicist colleagues brought along acoustic sounders, air guns and side-scan sonar systems among other tools for their investigations. Water depth and magnetic field data were collected while steaming between stations and along specific transects during night time hours after sediment and water sampling activities had ceased. Geophysical instruments also recorded changes in the strength of the earth's gravitational field along *Hudson's* track. Gravity data revealed information on the general characteristics of the earth's crust over which the ship was passing and on the more subtle effect that gravity has on sea level height from place to place. One set of gravity data from an 11, 265 km long transect that was run along longitude 150 west in the Pacific Ocean would ultimately assist in calibrating the altimeters on future orbiting satellites designed to measure small changes in sea level heights over broad areas of the world's oceans.

At the time of the expedition, the two samplers used by geologists for retrieving surficial seafloor sediment were the Dietz-Lafond sampler and the VanVeen sampler. The Dietz-Lafond sampler is the smaller of the two types. Its spring-loaded jaws are easily jammed in an open position by pebbles or larger rock fragments. The VanVeen sampler proved to be less susceptible to jamming and, in addition, was often able to recover a reasonably undisturbed sample of the thin, loosely consolidated uppermost layer of surficial sediment. Sediment samples were needed to evaluate water depth–related changes in sediment grain size and mineralogy, and to study the distribution of protozoan and invertebrate species whose skeletons and shells had accumulated in surficial sediment deposits.

Piston core sampling, on the other hand, was aimed at collecting 10 metre or longer 6 centimetre diameter tube-shaped samples of sediment. They were needed to study environment changes over a range of time intervals that varied in each core depending on the length of the core sample and the sedimentation rate at a particular location. Rock samples were collected by dragging a heavy dredge sampler over a hard rock surface to break smaller pieces of rock away from larger ones or to recover smaller rocks lying on the seafloor. Some of these originated from local deep sea bedrock outcrops while others were deposited by melting sea ice that had transported them from shallow near shore areas.

Some of the *Hudson 70* geologists were specialists in marine micropaleontology. They wanted to study the distribution of living planktic species that are found as fossils in ocean sediments. They could often be found tending to plankton samplers of one kind or another. Their main organism of interest during the expedition was a marine protozoan known as Foraminifera. The vertical plankton net used for this work during *Hudson 70* had a half metre square opening. Its frame was often fitted with a flow meter to allow an estimate to be made about the number of organisms per cubic metre of water collected by the net during a single tow. The net was typically lowered to a depth of 200 metres and then winched back to the surface. A more sophisticated design of plankton sampler called the Multiple Plankton Sampler (MPS) consisted of a set of three nets that could be opened and closed at various depths giving discrete plankton samples for successively deeper water layers. During the expedition, the MPS was deployed to depths of about 900 metres.

Unlike its vertically lowered counterpart, it was towed forward very slowly while being lowered through a range of water depths. Individual nets were opened and closed by pressure-activated releases as the sampler passed through deeper water depths during the lowering process. Information about the sampler's depth and other operational data was continuously sent back to the ship using a BIO-modified acoustic telemetry system that was interfaced with the sampler's sensors by BIO engineers.

By the time the expedition had finished, planktic Foraminifera-water mass linkages were defined for equatorial and northern Pacific waters. In addition, marine geologists had mapped the distribution of sediment types sampled in Arctic ocean basins, identified the occurrence of underwater pingos, and mapped ice scour patterns on the Beaufort Sea continental shelf using side-scan sonar. Their refraction seismic survey of Baffin Bay showed that its basin was underlain by oceanic rather than continental-type rocks thereby settling a question that, up to that time, had not been resolved.

Biologists from various government laboratories and educational institutions throughout Canada participated in many of the 9 legs of *Hudson 70*. Their areas of interest covered the full spectrum of marine life habitats—from sea-birds to sea mammals to fish to deep sea worms and everything in between. Three of their key collecting tools were the Isaacs-Kidd trawl, the epibenthic sled and the Smith-McIntyre sediment sampler. These devices were augmented by several plankton nets of various shapes and sizes.

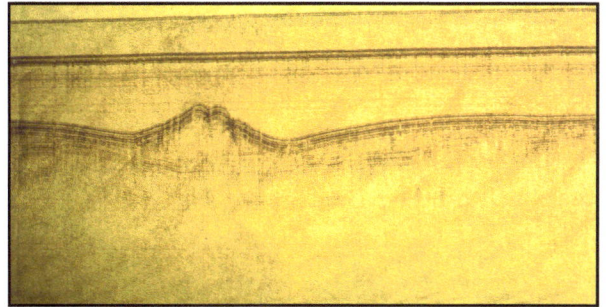

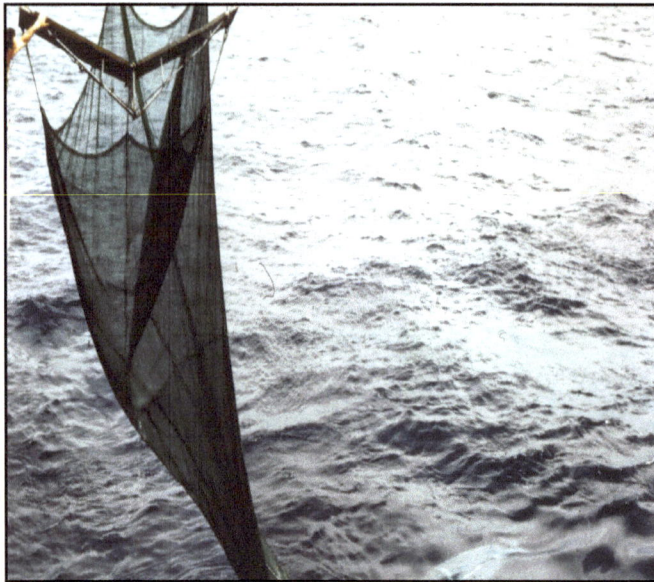

Indirect measurements of the density of plankton were made by staff from the Defense Research Establishment-Atlantic through experiments that measured the absorption and reflection of transmitted sound signals by the ocean's Deep Scattering Layer. During one of the southern legs of the voyage, biologists from the National Museum of Canada and the University of British Columbia were dropped off at the southern Chilean port of Puerto Williams for about a month so that they could study coastal and intertidal marine life. The research platform used for some of that work was *Hudson's* survey launch *Redhead*. Observations of seabirds, whales and other marine life were carried out almost continuously throughout the entire expedition.

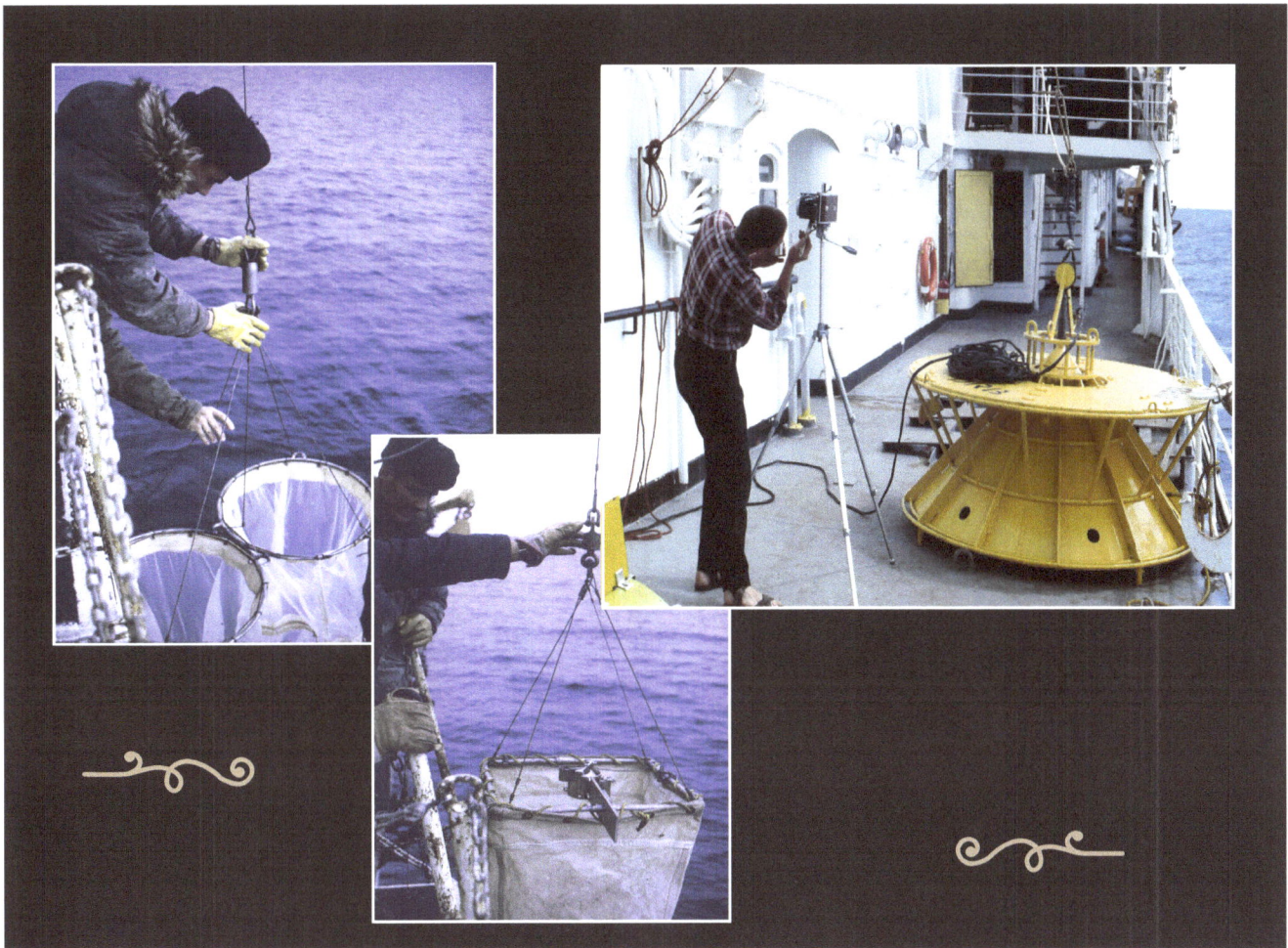

By way of comparison, a modern voyage of discovery such as the one carried out by *Hudson* in July of 2010 demonstrates just how much technology has changed the way that biologists are able to study sea life. The July 2010 expedition surveyed the submerged tops of Flemish Cap and Orphan Knoll off NE Newfoundland and the Sable Gully off Nova Scotia using an ROV (remotely controlled vehicle) fitted with various types of cameras and capable of working to depths of up to 3000 metres. Photos of seafloor life captured by the ROV's cameras were transmitted back to BIO in Dartmouth, NS., to a cultural centre in St. John's Newfoundland and to the Canadian Museum of Nature in Ottawa for study by scientists and viewing by the public. By the end of *Hudson 70*, biologists had sampled and accumulated a great deal of new information about species distributions in the air, in the water and on the seabed. Through their extensive sampling work, they documented South Atlantic pelagic ostracod occurrences between the equator and latitude 55°S and Pacific planktic Foraminifera from 10°S to 55°N. Those data represent a baseline that has potential for evaluating the impact of the late 1940s to 1972 cooling trend on these two organisms.

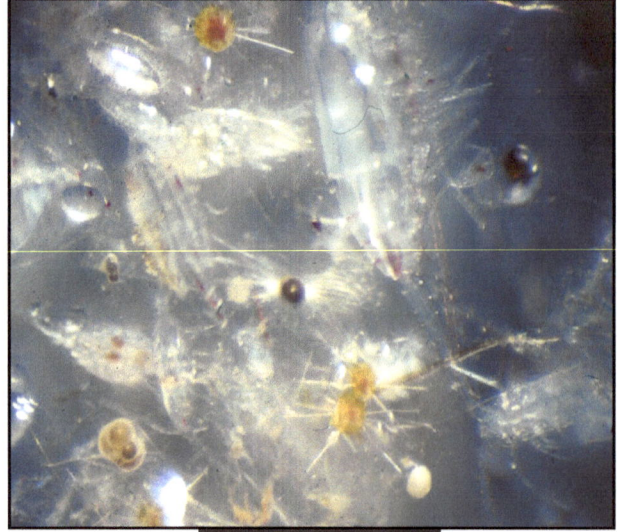

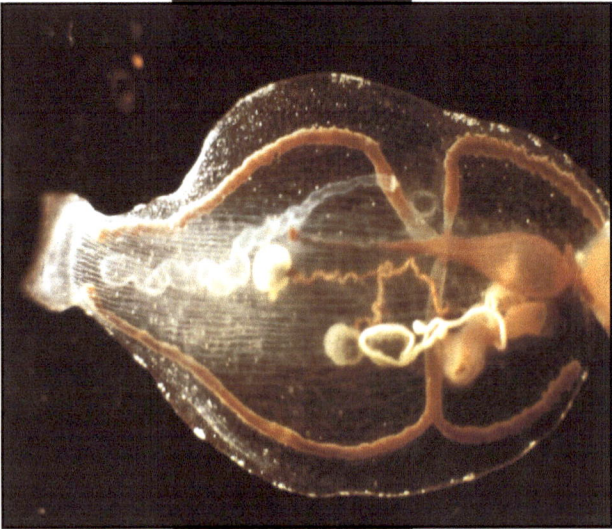

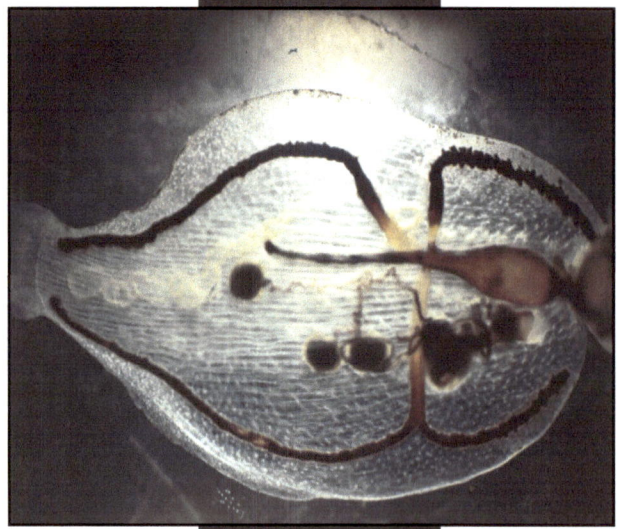

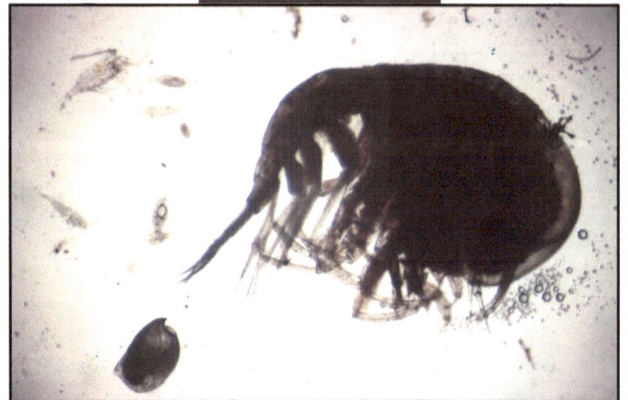

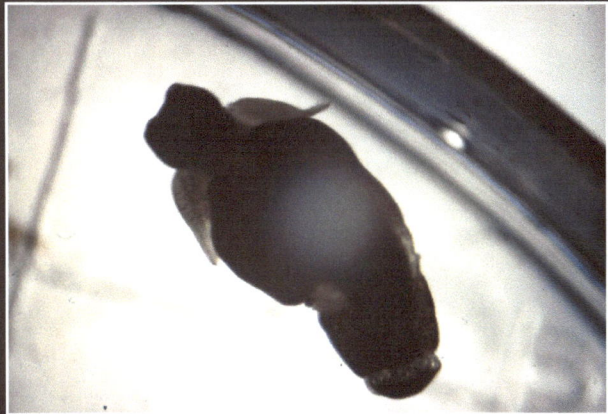

—Other Shipboard Observations and Tasks—

❧

What was seen from Hudson's Rail and in her Laboratories

Scientific observations made during the expedition were distinctly multidisciplinary and multifaceted. A few biologists were specifically interested in the distribution of seabirds in South Atlantic and sub-Antarctic climates while others watched diligently for sea mammals. Samples collected, using plankton nets, sediment samplers and an epibenthic sled, were processed on board to allow preliminary assessment of their marine life forms. They were then preserved and packed for later study at various laboratories throughout Canada.

Oceanographic technician's efforts focused mainly on oxygen, salinity and nutrient analyses of water collected from bottle casts. Biologists also collected water samples, some of which were unusually large, to evaluate the particulate organic matter content of seawater. Geologists were responsible for the logging of sediment core samples and for the preliminary description of sediment grab samples. Geologists that were specialists in marine micropaleontology, started mapping the regional distributions of planktic and benthic species observed in water and sediment samples to develop ideas for descriptive models needed to interpret occurrences of their fossil counterparts in ancient marine sediments. Superimposed on all of this *hands on* activity was the monitoring of instruments and chart recorders that displayed remotely sensed information about the geometry of seafloor sediment deposits and changes in the Earth's gravity and magnetic field along the ship's track. This is the domain of geophysicists that specialize in field methods that encompass a variety of remote sensing technologies.

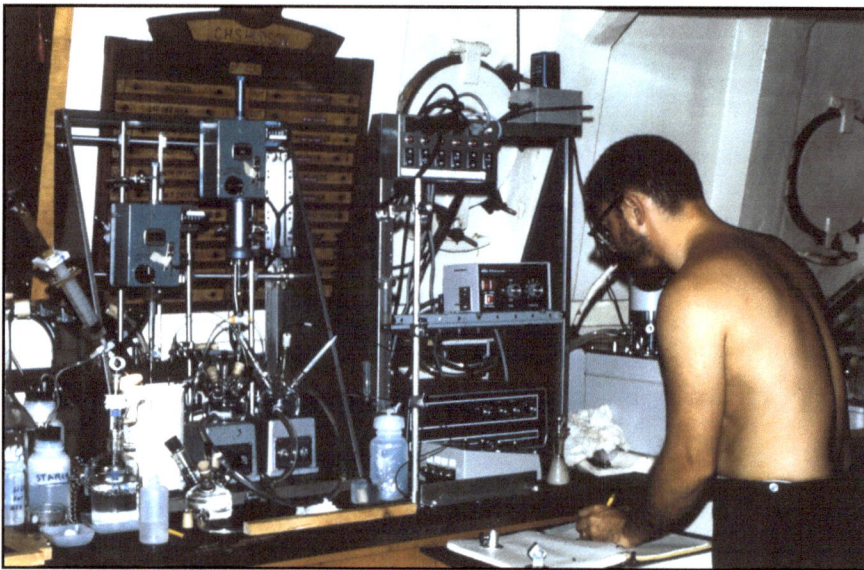

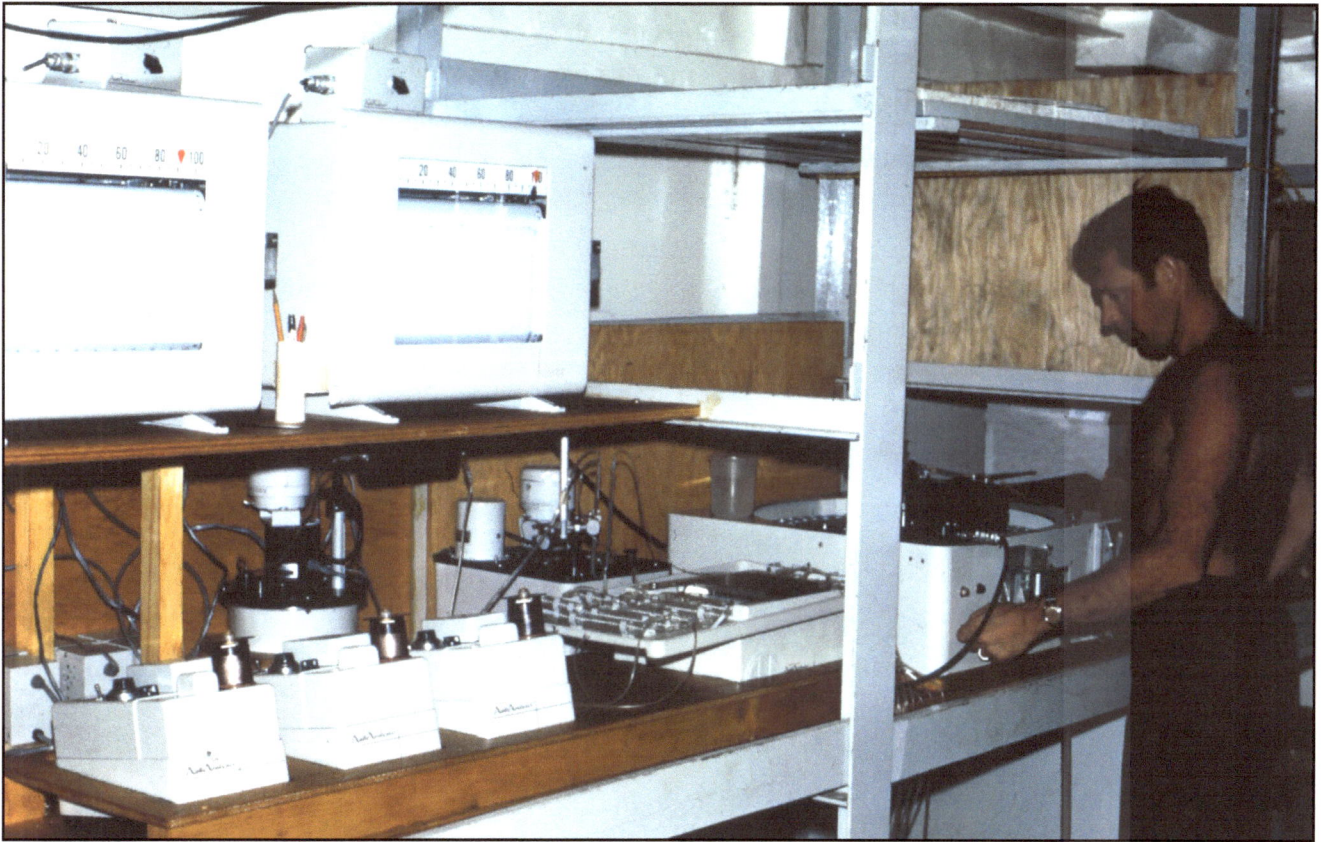

Their reflection seismic and side-scan sonar survey systems are based on the use of sound to create echoes that paint a picture of seafloor surface features, or of the internal structure of thick marine deposits. These systems were important elements of the *Hudson 70* geophysicist's tool kit. Other instruments made complementary measurements of various geophysical parameters such as changes of the Earth's gravity and magnetic fields from place to place. For some of these measurements, sensors were often encapsulated in waterproof tubular-shaped cases or *fish* and towed behind the ship between sampling stations that, in some cases, were more than 600 kilometers apart. Following the expedition, geophysical data were evaluated together with sediment grab and core sample results to provide a comprehensive picture of the seafloor under investigation. This strategy was particularly effective in studies of several areas of the Arctic Ocean basin surveyed during the voyage.

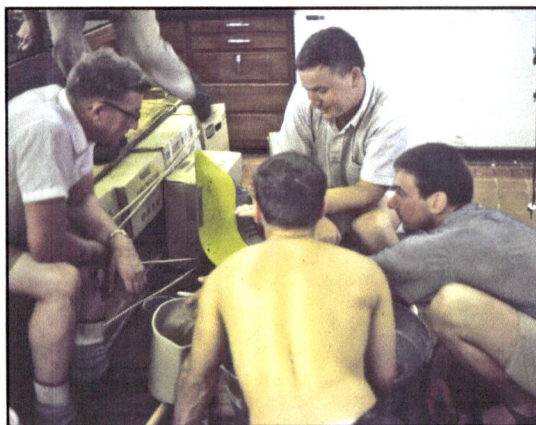

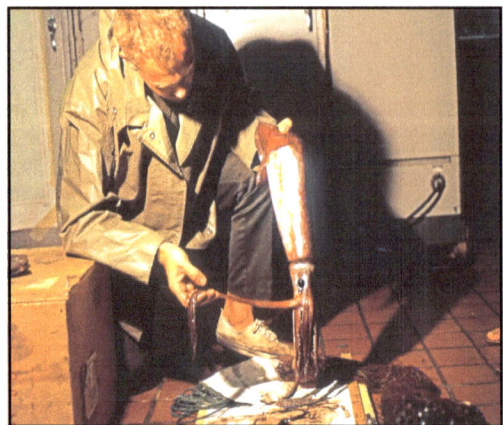

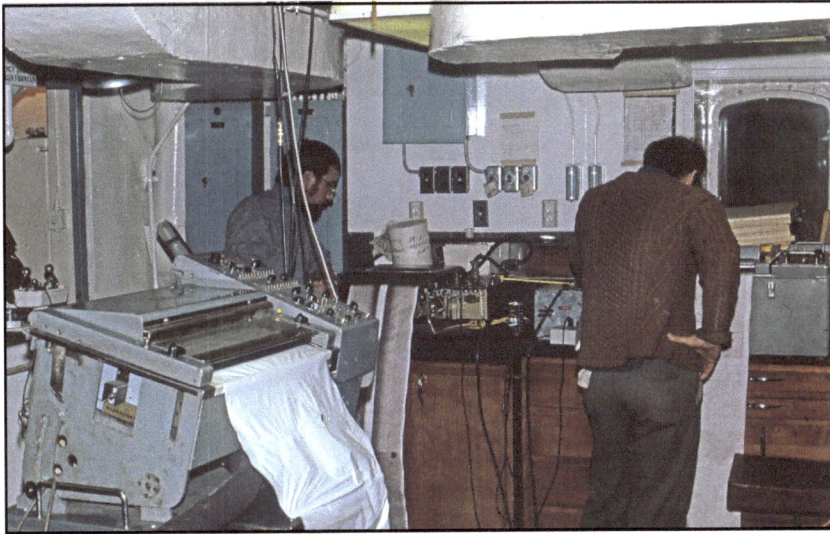

Sediments collected during the expedition, using grab samplers and rock dredges, provided many clues about the depositional environment from which they were retrieved. These data were typically derived from analyses of sediment grain size distributions, mineral content, chemical constituents, and from the sediment sample's microfossil content.

— Shipboard Routines —

SAFETY, WORK
AND RECREATIONAL BREAKS

Day-to-day activities of *Hudson 70* participants focused on two broad goals, namely, completing the scientific program and returning the staff and crew safely to each port of call and finally to Halifax. The suite of support work needed to realize these two goals involved an ongoing, 24/7 effort by scientists, technicians and, especially, by *Hudson's* officers and crew.

Safety and lifeboat drills were first and foremost for this kind of multi-leg expedition. They were repeated and practiced often during the voyage in order to familiarize new staff and crew that joined the ship at various ports with emergency and other protocols that were to be followed. Lifeboat deployment, operation and equipment familiarization exercises were kept in the forefront by using the lifeboats for both safety drills and for various scientific tasks. Lifeboats were often deployed to transport geologists and biologists to shallow nearshore areas of interest that were otherwise inaccessible by *Hudson* or by her launches.

Daily operations on *Hudson* were typically tedious and repetitious. Many of the daily tasks involved lowering and raising instruments and samplers on very long lengths of thin wire cable. This work was done with several winches that were specifically designed for that kind of operation.

Other deployments and recoveries were done using *Hudson's* lifting cranes. One especially difficult operation was the deployment and recovery of current meter moorings. That task required coordination between scientists and technicians along with members of *Hudson's* deck crew and her officers working at various tasks on the ship's bridge. The retrieval of either an instrument such as an underwater camera or a current meter mooring generated hours of work for technical staff that were responsible for processing film or transcribing current meter data tapes. Geologists and biologists that were fortunate in having recovered large samples of sediment or marine life worked long hours sub-sampling, labeling, processing, preserving, identifying specimens, and logging sediment cores and grab samples. Much of this work was done as the ship steamed from one station to the next, sometimes traveling for several days especially during the expedition's long mid- ocean legs.

By the time *Hudson* reached the Beaufort Sea area of the western Arctic, the scientific program called for a number of closely spaced survey lines. As a consequence, the routine changed to one in which geophysicists towed their gear during daylight hours while geologists, oceanographers and biologists deployed and retrieved samplers, sensors and underwater cameras and television systems during lower light evening hours.

Hudson's deck crews were in constant motion, both during deployment and recovery operations, and in preparing, maintaining and reorganizing the ship's equipment for the next suite of scientific activities. All scientific mission-related tasks were superimposed on a list of ongoing ship-related maintenance and repair work aimed at keeping *Hudson* operating safely and at maximum optimal efficiency throughout its 11 month voyage. Galley staff seemed to be on their toes every minute preparing meals and snacks for the entire ship's complement that included the occasional surprise dish.

Daily round-the-clock activities of *Hudson's* staff and crew were interlaced with well-earned rest and recreation sessions. They varied from typically short intervals of deep sleep in a bunk that was linked to the changing motions of the ship under various sea state conditions, to more relaxing pursuits. *Hudson's* two lounges were popular gathering places for ship's crew and staff. The officer's lounge was often filled with scientific staff that had just completed their work shift and were eager to share their day's experiences and discoveries with anyone that would listen, or to just catch up on the latest news from around the world.

Communication between staff and officers was catalyzed by the presence of a bar facility that offered a limited supply of beer and other alcoholic beverages at duty-free prices. On sunny and warm weather days, many of the off duty staff could be found on *Hudson's* upper deck. This area was also very popular among joggers and power walkers that found themselves in need of regular physical exercise over and above that encountered during their lengthy daily routines. Others just seemed to enjoy a spectacular sunset.

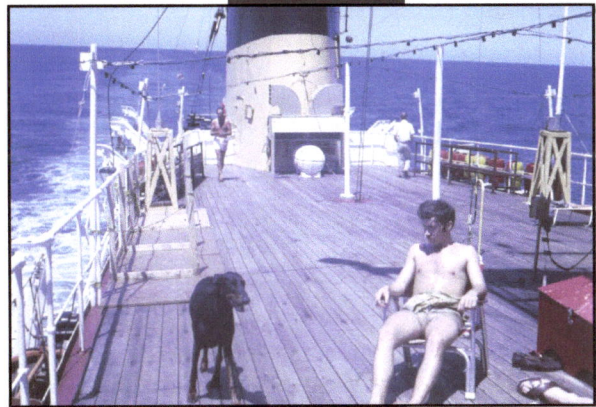

— Other Participating Ships —

Sometimes it takes Teamwork

It would be ideal if research ships were more versatile and able to carry out whatever scientific and survey tasks were needed for particular programs. However, designing a research ship to meet all the requirements of Canada's rich and diverse marine sciences programs would not be practical, cost-effective, or even possible, given the range of ocean and coastal environments in which they must be able to operate. So, from time-to-time, we find *Hudson* joining forces with several other ships to accomplish a scientific or survey task, or to be able to access ice-covered ocean areas that were beyond her own ice-breaking capabilities.

The first of the *Hudson 70* multi-ship operations involving a larger vessel occurred off the coast of British Columbia in July. It consisted of a two ship seismic refraction profiling survey of a part of the continental margin. That effort contributed new knowledge about plate tectonics in this part of the eastern Pacific. *Hudson's* partner for the geophysical work was the *CNAV Endeavour*. Her mission also included rock dredge sampling at underwater places with names like Delwood Knolls and Paul Revere Ridge.

During the refraction seismic survey, *Endeavour* served as the shooting ship *i.e.*, launching and setting off underwater explosive charges or *shots*. The sound waves generated by the explosions were recorded by *Hudson's* streamed array of hydrophones *i.e.*, by the listening ship. The two ship operation resulted in the successful completion of two reverse refraction seismic lines, each about 91 kilometers long. They required detonating a total of 201 *shots* ranging in size from 23 to 272 kilograms.

CNAV Endeavour was launched in September, 1965. She was 72 metres long and displaced 1560 tonnes. With a complement of 50 officers, crew and scientists, this intermediate size research vessel was especially effective for inner continental shelf and coastal missions. *Endeavour* was decommissioned in 1998.

CSS Richardson, on the other hand, was a considerably smaller vessel that was most suited for working in bays, estuaries and inner continental shelf environments under appropriate weather conditions. As such, her almost 19 metre length and 59 tonne displacement was ideal for Arctic coastal survey work that might be problematic for larger vessels such as *Hudson* - especially in the relatively poorly charted waters of the Beaufort Sea. *Richardson* was launched in March, 1962 and decommissioned in 1990 after 28 years of service for various federal government departments and the Canadian Hydrographic Service. During the summer of 1970, she preceded *Hudson* to the Beaufort Sea where the on-board team of geophysicists carried out sounding, reflection seismic and side scan sonar surveys in coastal zone areas such as MacKenzie Bay. Following the arrival of the *Hudson*, some of *Richardson's* geophysical team transferred to the larger vessel in order to extend their surveys to outer continental shelf and deep ocean basin environments.

Near the end of the summer of 1970, *Hudson* was joined by the *CSS Baffin* and the Canadian icebreaker *Sir John A. MacDonald*. *CSS Baffin* was built in 1956 by Canadian Vickers and was Ice Class I certified. At the time, she was considered to be one of the most modern hydrographic survey vessels in the world with a length of 90 metres and a displacement of 4986 tonnes.

Although comparable in size to *Hudson*, the *Baffin* was outfitted specifically for hydrographic work. She could carry up to six hydrographic launches and two landing barges and was also designed to accommodate two helicopters. Her launches were particularly well suited for charting shallow in-shore areas of Arctic coasts. With a complement of 57 officers and crew, and accommodations for up to 29 hydrographers, *Baffin* was initially deployed to the Arctic in the late 1950's and from time-to-time in the 1960's and 1970's to carry out hydrographic surveys in places such as Lancaster Sound, Maxwell Bay, the Baffin Bay offshore, *Hudson* Strait, Southampton Island and, in 1970, in the Beaufort Sea.

Both *Baffin* and *Hudson* were fortunate in having the icebreaker *Sir John A. MacDonald* on hand in the summer of 1970 for logistical support, especially in the often ice-choked channels of the eastern Arctic archipelago. The *MacDonald* displaced 9000 tonnes or almost six times that of the *CNAV Endeavour*. Commissioned in 1960, the *Sir John A.* served Canada's icebreaking needs for almost three decades. During *Hudson 70* operations, she provided the open water conditions required by *Hudson* and *Baffin* to complete the oceanographic, geophysical and depth sounding programs that marked all of the earlier southern legs of *Hudson's* journey of discovery. More importantly, the *Sir John A.* represented the critical logistical support that would ensure that *Hudson* reached the ice-free waters of Baffin Bay and then eventually go on from there to complete her circumnavigation of the Americas. Built in Lauzon, Quebec in 1960, the *MacDonald* was decommissioned in 1991 and sent to the scrap yard in 1994.

There was one other ship that had a critical supporting role in the *Hudson 70* geophysical program carried out in Baffin Bay. Like the *Endeavour*, the United States Coast Guard ship *Edisto* would serve as the shooting ship in a refraction seismic survey of Baffin Bay that would confirm the oceanic crustal nature of the Bay's underlying basin. Commissioned by the U.S. Navy in March, 1947, and classified as a Wind Class icebreaker, *Edisto*, at 3575 tonnes displacement and 82 metres in length, was 8 metres shorter and about 1400 tonnes lighter than *Baffin*. However, its internal infrastructure was designed to accommodate a total complement of 205 personnel as opposed to *Baffin's* normal complement of 86. During the *Hudson 70* Baffin Bay seismic survey, *Edisto* set off 54 explosive charges. She was recommissioned by the U.S. Coast Guard in October, 1965 and served in various roles until decommissioned and scrapped in November, 1974.

— WHAT WAS DISCOVERED —

SOME EXAMPLES OF WHAT WAS ADDED TO THE GLOBAL POOL OF KNOWLEDGE

The years following the *Hudson 70* expedition witnessed the publication of a total of at least 45 scientific papers, atlases, data reports or magazine articles based on the diverse suit of investigations and surveys that were completed during the voyage. New papers continued to appear in the scientific literature as late as the early 1990's. The expedition's scientific publications can be broadly categorized as either baseline studies, locally targeted investigations or as experimental and methodological. They can be further subdivided into oceanographic, biologic and marine geology/marine geo-physics subcategories. Many of the locally targeted studies were, in effect, baseline in nature because the work was conducted in particular parts of ocean basins that likely had never been visited by a research vessel but that were now accurately positioned using satellite navigation.

Atlantic Water Masses (Dissolved Oxygen)

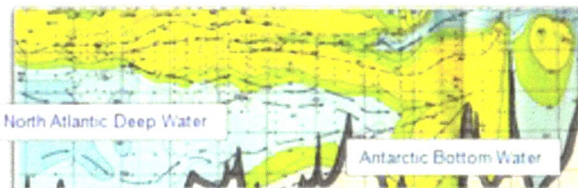

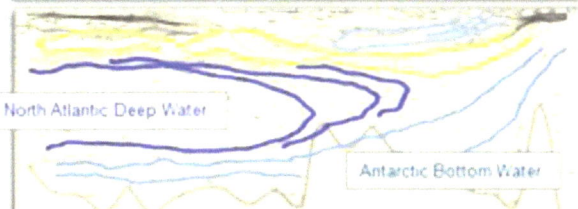

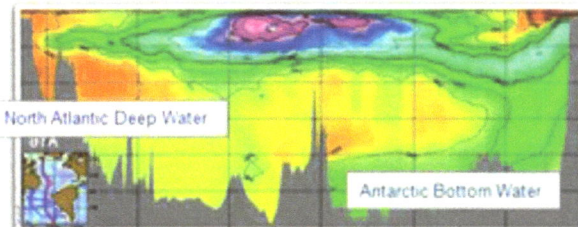

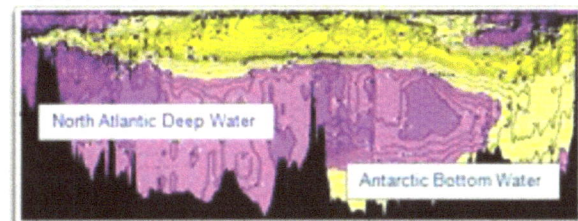

- German Meteor Expedition 1925 – 1927
- Hudson 70 Expedition 1969 - 1970
- GEOSECS Expedition 1973 – 1976
- WOCE Expedition 1988 - 1998

9/12/2010   Hudson 70 Expedition - Dr. Iver Duedall

The two dimensional mapping of water mass characteristics along *Hudson 70*'s mid-ocean transects provide a 1970 footprint of the patterns of South Atlantic and Pacific ocean water masses that reflect a variety of mixing processes. For decades, both nutrients and dissolved oxygen were used to define and trace oceanic water masses. These two sea water properties, in addition to salinity and temperature, facilitated a quantitative analysis of the ocean's circulation. When combined with similar measurements taken at different times by other voyages, *Hudson 70* results contribute to theories about how the circulation behavior of the oceans might be expected to respond to longer-term climate variations. As such, *Hudson 70*, as well as past and future expeditions, provide the essential baseline information that is needed by scientists to improve understandings about the way that the ocean system reacts to the warming and cooling trends that drive its well known conveyor belt-style circulation system.

The impact of climate on the oceans is, however, not all about water mass characteristics. Distribution patterns of various life forms in space and time offer another strategy for assessing the effect of global-scale processes such as climate change, or of more local effects that are, for example, perhaps linked to man-made events such as oil spills or to longer-term contamination by various chemical compounds and trace metals. For example, one of the surveys carried out during the Tahiti to Vancouver leg of *Hudson 70* was aimed at collecting information about the distribution of planktic Foraminifera that live in the upper 1000 metres of the water column. These marine protozoan's have a skeleton made of calcium carbonate that's about the size of a small sand grain. The various species of planktic Foraminifera are distributed in the ocean mostly according to their water temperature and salinity preferences. Understanding the links between water mass characteristics and their associated populations of planktic Foraminifera offered another strategy for investigating a number of contemporary environmental issues.

When planktic Foraminifera die, their skeletons sink to the seafloor and eventually become buried by marine sediment. Scientists interested in long-term climate change can often be found taking core samples that may record (depending on the length of the core and the sedimentation rate) thousands of years of water mass variations. By comparing the fossil assemblages of Foraminifera preserved in a core sample with the known distributions of living species at various times and conditions, scientists can develop theories about long-term climate changes that have modified water mass features. Samples of living planktic Foraminifera collected during the Tahiti-Vancouver leg are of particular value because they reflect conditions in the mid-Pacific ocean that existed near the end of the mid–20th century cooling trend. The *Hudson 70* survey identified five water temperature-dependent groups of species in three major Pacific water masses. It also revealed the details of a northward decrease in species diversity.

The Arctic leg of *Hudson 70* began on August 14th when the ship departed from Victoria, B.C. As she steamed north toward the Bering Strait, her water depth recorders revealed rectilinear fault patterns and rugged topography. She eventually arrived at Herschel Island where the Beaufort Sea survey would begin. The marine geology and geophysical survey of the Beaufort Sea was the major scientific activity carried out during this phase of the expedition. Other ongoing work continued to provide a comprehensive data set of the Sea's life and water mass characteristics.

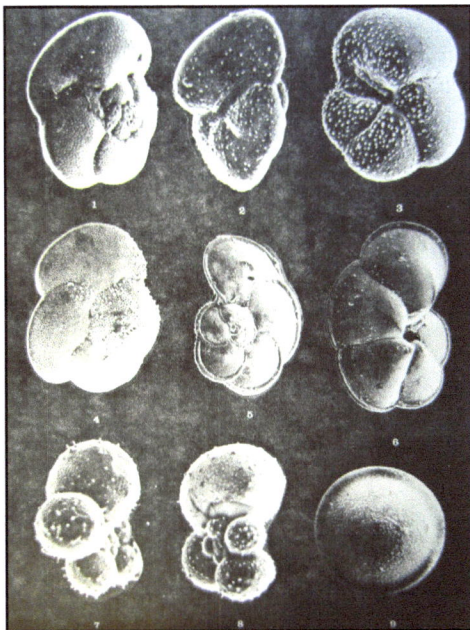

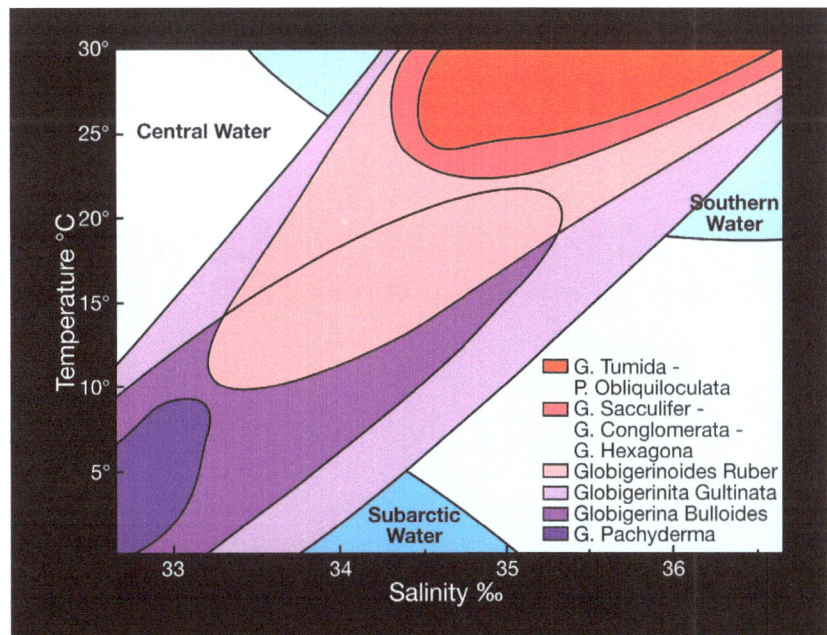

PINGOS:
Coastland and offshore zones of pingos

BEAUFORT SEA

cold Arctic water (salinity = 33‰ and temperature = -1.9°C)

warm Mackenzie water (salinity = 3 - 30‰ and temperature = 0° - 9°C)

Baillie Islands

Russell Inlet

Liverpool Bay

Peninsula

Tuktoyaktuk

Kugmallit Bay

Tuktoyaktuk

Eskimo Lakes

Mackenzie Bay

Anderson River

YUKON T.
N.W.T.

scale 1:1 000 000

LEGEND

Undersea

Area of known undersea pingos

Area of submerged bars, headlands and coastal pingos

Pingos
Landward boundary of known undersea pingos .... — LB —
Coastal circulation .... →

Coastland

Pingos ....
Southern boundary of land pingos .... — B —

25   0   25
kilometres

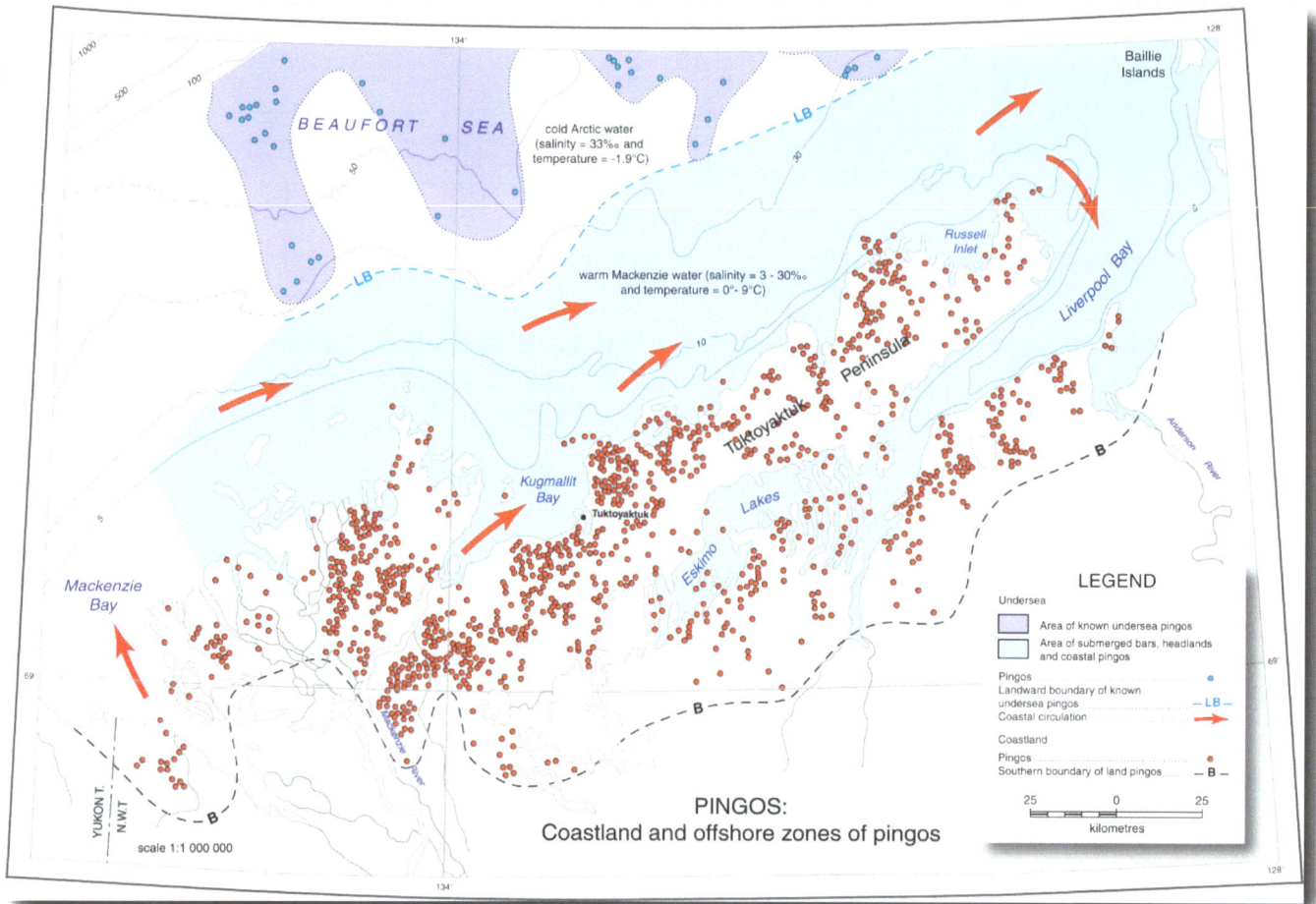

Abrupt local shoaling of the ocean floor in the Beaufort Sea was first noticed in 1969 by hydrographers aboard the *John A. MacDonald* while escorting the tanker *S.S. Manhattan* to Prudhoe Bay, Alaska. *Hudson* and Baffin would go on to discover dozens of these suspected ice cored submarine mounds or pingos (see the 1988 atlas published by B. Pelletier). At mid-shelf depths, *Hudson's* side scan sonar equipment gathered evidence of ice-scouring while biologist's plankton net tows showed the widespread absence of many eastern Arctic planktic species. Scientist's seafloor sampling work was gradually painting a picture of linkages between life on the ocean bottom in relation to sediment texture and seabed morphology.

Seismicity

modified from Weaver and Shedlock, 1996

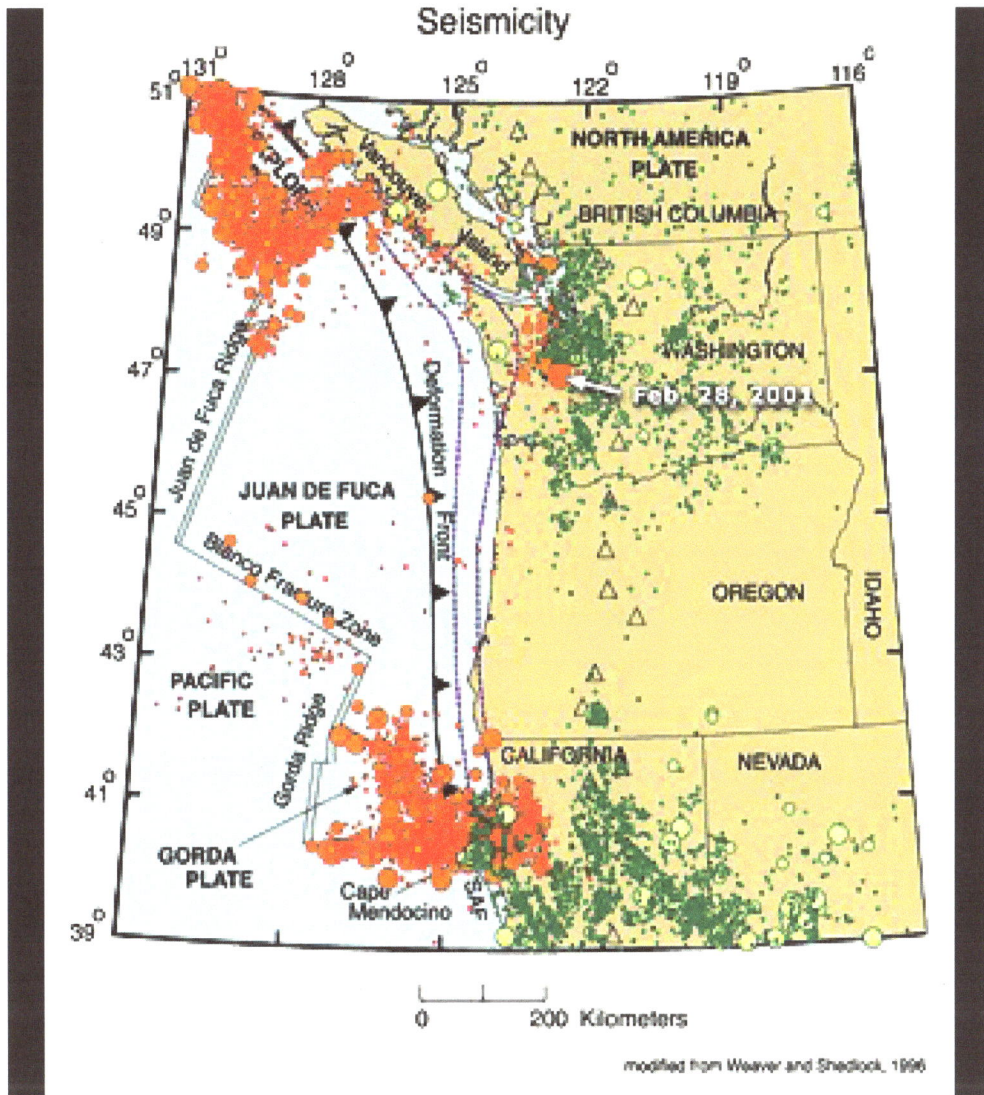

The geophysical survey of the edges of the Juan de Fuca tectonic plate off Canada's west coast was among the targeted studies of *Hudson's* scientific program. One heat flow measurement made during that survey became the record for heat flow measurements made at sea in the era before the discovery of hydrothermal vents. At the time, *Hudson's* geophysicists were hoping to add new information that would help in understanding plate tectonics theories that are still being researched and debated today.

However, the more immediate aim of the survey was to apply the geophysical techniques developed at BIO and elsewhere to the region of the ocean floor lying west of the southern part of the Queen Charlotte Islands and Queen Charlotte Sound. That goal placed *Hudson* over the northern termination of the Juan de Fuca Ridge and the area surrounding the Explorer Trench.

Preliminary analysis of geophysical data from the survey showed the presence of a sedimentary basin at the foot of the continental slope in which the sediments are progressively more intensely deformed from north to south. This feature was interpreted as an indication of the interactions between adjacent lithospheric plates. Other observations pointed to Explorer Ridge as a presently active spreading segment of the East Pacific Rise. As you might expect, research on the processes behind lithospheric plate movements has continued to the present day. Recent research is producing new explanations as to why tectonic plates move along the Earth's surface at the speeds that are being observed. Surveys such as the one completed by *Hudson 70* scientists in the Juan de Fuca area have contributed to the field observations that underlie hypotheses about this major Earth process.

Marine specimen collecting and field observations of birds and marine mammals was an ongoing effort throughout the entire expedition. Information gathered over the 11 month voyage yielded several new species. However, more importantly, it accumulated data that could be used in future surveys such as the recently completed Census of Marine Life international project. The first summary of Census results were released in 2010 but the idea for it goes back to questions about marine species diversity that were being posed in the late 1990's. Archival data examined by Census staff during that decade-long effort included observations of the type reported by *Hudson 70* biologists.

However, the Census project itself was of a much larger scale, involving as many as 540 expeditions to all ocean realms and collaborations between 2700 scientists from 80 nations. Information of the kind brought together by the Census project has allowed comprehensive baselines to be drawn that have utility for measuring the impacts of natural changes, and of human actions, on ocean environments. Similarly, the 11 month-long effort of *Hudson 70* biologists, produced new information on (i) distributions of sea birds in the Drake Passage area and off the coast of Argentina, (ii) distribution of South Atlantic pelagic ostracod species, (iii) relationships between Pacific planktic Foraminifera and their host water masses, (iv) a new species of lantern fish (*Diaphus hudsoni*), (v) the post-larva stage of some stomatopod crustacean species, and on (vi) brood protection in a species of epipelagic pteropod.

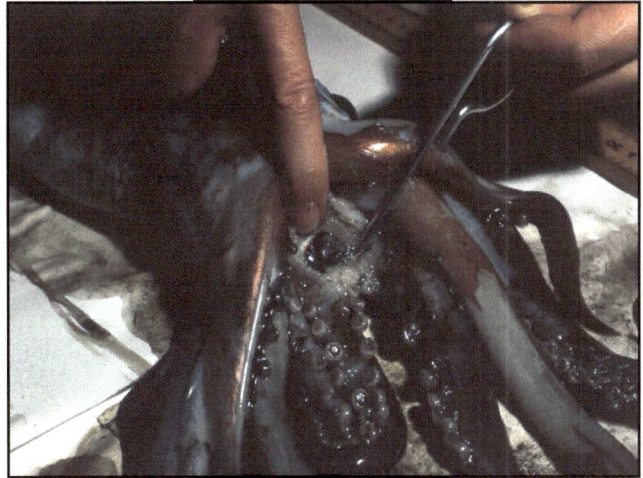

New publications that rely on *Hudson 70* data and archived samples can be expected from time-to-time for many years to come. Unfortunately, the scale, scope and duration of this unique voyage of discovery was probably a once-in-a-lifetime Canadian ocean sciences event that will likely not be repeated anytime soon.

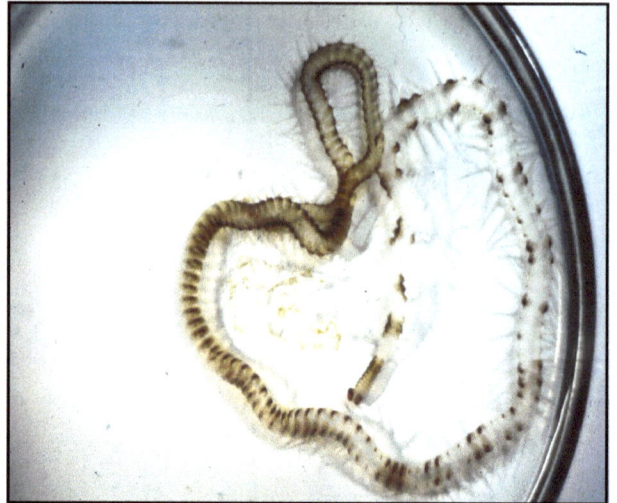

— Celebrating Hudson 70

After 40 Years —

A Surprising number of Expedition Participants Showed up

More than forty-two years ago, at 0800 on September 30, 1970, a shore party of three *Hudson 70* participants climbed a small hill on Cornwallis Island near Resolute and anchored a plaque to a bedrock outcrop to commemorate *Hudson's* arrival at the eastern end of the Northwest Passage. Boatswain Joe Avery, marine photographer Roger Bellanger and Chief Scientist Dr. Bernard Pelletier seemed to have an inkling that something unusually profound was taking place that would, among other accomplishments, set a new record for a Canadian research vessel—the first circumnavigation of North and South America .

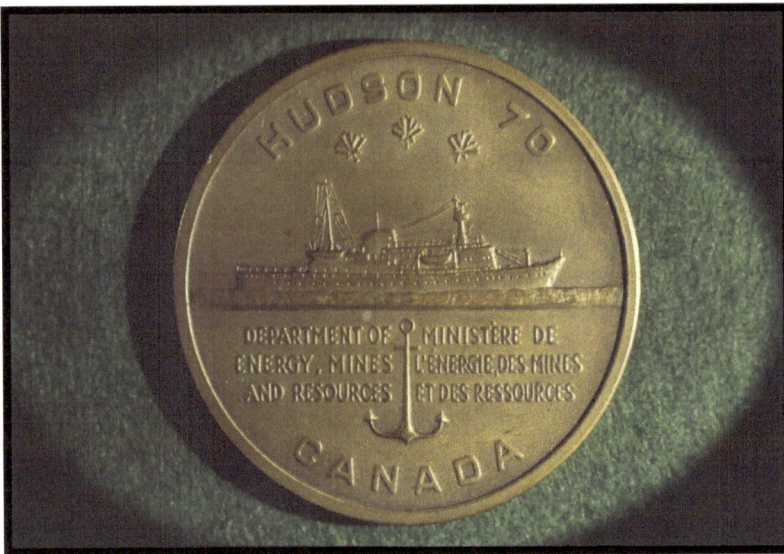

In August of 2002, Mr. George Fowler of BIO's Ocean Physics and Technical Services Section and several of his coworkers were on a port call at Resolute and decided to relocate the plaque that had been left there 32 years earlier. They found it at a site 30 metres above sea level on a rock outcrop at Cape Martyr. Fowler reported that the plaque was in remarkably good condition except for a few bullet holes.

In early 2009, BIO announced its plan to celebrate the 40th anniversary of the *Hudson 70* expedition on November 17th and 18th of that year. At that point, it seemed as though no one realized that memories of the voyage would attract as many as 50 of the original participants. The idea for the celebration was sparked by two of the expedition's youngest scientific staff, one of whom was on board for the entire voyage. At the time, Roger Smith and Peter Wadhams were university students that elected to take breaks from their studies in order to participate in the voyage. For stories about daily life during the voyage see Wadhams (2009).

On the morning of the first day of the two-day gathering, the participants filed in to the BIO auditorium and were seated. They and the attending audience were welcomed by Mr. Tom Sephton. Next, *Hudson* participant Roger Smith introduced three departmental representatives that each made a short presentation on the significance of the expedition in relation to BIO's history. Then each of the voyage's participants was called to the front of the auditorium to be recognized and to receive, among other memorabilia, a commemorative coin and a commemorative pen made using teak wood that was cut from a section of *Hudson's* guard rail. The individual presentations were followed by a song The Mighty *Hudson* (written and sung by two current BIO staff) after which Mr. Ed Murray described two time capsules that had been prepared by several members of the *Hudson 70* Organizing Committee. One of the capsules is being carefully archived at the BIO library and the other has made its way back to Resolute where it is to be displayed at a public place with an associated poster explaining its significance. The BIO library capsule will be opened at the 100th anniversary of the Institute in 2062. After the auditorium part of the ceremony was completed, the participants and their family members walked down to the BIO jetty and boarded *Hudson* to view a series of exhibits of early 1970's research tools and several video presentations of operations and survey activities that had taken place during various legs of the expedition. The luncheon on board *Hudson* that followed the exhibition tour was sponsored by the Canadian Coast Guard which is presently responsible for the maintenance and operation of the ship. At about 3:30 PM, participants and family members, along with invited guests and VIP's, reassembled in the BIO cafeteria for a home made clam chowder and wine tasting reception where old friendships between shipmates that, in some cases, had strayed over the decades were renewed.

The second day of the celebration was reserved for a *Hudson 70* science symposium in which the local public and scientific communities were offered a series of talks by some of the expedition's participants. Presentations covered several of the voyage's key investigative activities in oceanography, marine biology, marine geology, and marine geophysics. The list of speakers included Dr. Bernard Pelletier, the oldest surviving member of the *Hudson 70* scientific staff at a lively 86 years of age. The last talk of the afternoon was given by one of the two young graduate students that had participated in the entire voyage. In his concluding remarks, Professor Peter Wadhams reminded the audience that the *Hudson 70* expedition *was both the first circumnavigation of the Americas and the last of the big multidisciplinary oceanographic expeditions that hacked out our basic knowledge of our ocean's structure, water masses and currents during the 19th and 20th centuries.* Since then, literally hundreds of shorter targeted surveys sponsored by the world's maritime nations have continued to fill in the gaps left by the earlier voyages of discovery. In an article describing an interview that was published in a 1971 issue of the Canadian Geographical Journal, Dr. Bill Cameron whom, at the time, was Head of the Marine Sciences Branch of the former Department of Energy, Mines and Resources, succinctly summarized Canada's rationale for the expedition. He remarked that *Canada cannot afford, for the preservation of her own scientific vitality, to restrict her efforts to purely domestic [marine] affairs... A country as great and as ingenious as ours has an obligation to contribute to basic knowledge of the world's oceans as a whole.*

Those thoughts reflect an attitude of confidence and proactive thinking that we don't seem to hear enough of these days. However, they also show a deep understanding of the true potential of Canadians and our responsibility as a nation to contribute to the pool of knowledge about the nature and status of all of the Earth's environments both above and below the low tide mark.

— THE HUDSON —

49 Years Old and Still Ready for More

A relentless workhorse and the pride of Canada's fleet of marine research vessels, *Hudson* remains a prime candidate for many Canadian marine investigators that are seeking to explore, measure and map the nation's off-shore environments. Formerly the *CSS Hudson*, she became the *CCGS Hudson* in 1996 and is now recognized by her coat of the Canadian Coast Guard's characteristic cherry red paint. Designed by Gilmore, German and Milne of Montreal, and built in 1963 by St. John Shipbuilding and Dry Dock Ltd. of Saint John New Brunswick at a cost of 7.5 million dollars, she is scheduled to be replaced in 2014 which would represent just over 50 years of service to Canada. Preliminary plans call for her replacement to be 90 metres long and to carry a team of up to 31 scientific staff. Given the real rate of inflation over the past several decades, it would not be surprising if her replacement's cost turns out to be several times that of her original price.

Hudson's displacement weight is 3444 tons. She is 83 metres long with a 15 metre beam and is powered by four diesel generators that drive the electric motors that are linked to her two propellers. Over the decades, *Hudson* has been the preferred platform for many of Canada's national and international deep sea expeditions. However, in addition to her research-focused specifications and on-board facilities, much of her success reflects the enduring performance and dedication of her officers and crew. Her design includes space for a complement of 37 scientists and she has a range of approximately 24,000 kilometeres. Those attributes have allowed *Hudson* to satisfy the logistical requirements of government and university scientists for operations in both open ocean and ice-infested waters. Among her earlier missions in the years prior to the *Hudson 70* expedition, the ship was used to study and sample the rock veneer and geophysical characteristics of Mid-Atlantic Ridge (MAR) mountains and valleys near 45 degrees north latitude. Underwater photography, seafloor sediment sampling and hard rock drill sampling along with bathymetric, gravity and magnetic remote sensing mapping activities over the crest of the MAR would provide the experience and techniques needed by scientists, officers and crew for many of the tasks carried out during the *Hudson 70* expedition when she was under the command of Captain David Butler.

A history of *Hudson's* career highlights between 1963 and 2002 was published by one of her current captains (Richard Smith) in the 2002 BIO Annual Review. They include, among other accomplishments, the first marine geology survey of the eastern part of the Grand Banks in early 1980, and an oceanographic expedition to Greenland in 1993 where the vessel bumped into a large iceberg that left a 4.6 metre long gash in her starboard side. In 1997 and 1998, *Hudson* was tasked for a major hydro-graphic survey mission to chart Rankin and Chesterfield inlets in Hudson Bay. The latter part of the last century finds her assisting with the Swissair disaster off Peggy's Cove Nova Scotia and, by the turn of the century, she was being used for surveys of the Gully area of the Nova Scotia continental slope near Sable Island.

Hudson continued in her marine research support role for various investigations and surveys during the course of the first decade of the 21st century. Between 2003 and 2010 we find her involved in a broad spectrum of multidisciplinary work in Canadian waters in support of both national and international research programs. BIO scientific staff, often in collaboration with scientists from various Canadian universities, have taken *Hudson* to many east coast and eastern Arctic marine areas for surveys and research. Those efforts have contributed to the sustainable development and responsible exploitation of Canada's inventory of offshore renewable and non-renewable resources. Her geological and geophysical missions have emphasized studies of benthic species habitat characterizations using remote sensing tools such as reflection seismic, side-scan sonar and multi-beam mapping systems. Similar research methods were used to assess foundation conditions and geohazards of continental slope and rise settings, for the geophysical delineation of submarine gas hydrate reserves, and to add to Canada's knowledge of regional sediment transport processes and bedform stability in commercially valuable areas such as the Grand Banks.

Meanwhile, oceanographers have used the ship to monitor water mass conditions in the Gulf of St. Lawrence and St. Lawrence Estuary, for regional monitoring of seasonal climate variability features of Scotian, Labrador and Newfoundland shelf waters, and to study oceanographic factors affecting the safe pursuit of offshore oil and gas exploration. BIO and university-based biologists have been equally ambitious in deploying *Hudson* for investigating and monitoring impacts of hydraulic clam dredging, the effects of drilling waste discharges on benthic communities, and for regional benthic habitat and deep sea coral distribution surveys. Virtually all of these investigations were aimed at increasing understandings of benthic marine community behaviour in both biologically important as well as in unexplored parts of ocean environments that fall within Canadian jurisdiction.

As we prepared the draft of this final chapter, there was a mutual feeling that we shall be hearing quite a bit more about *Hudson's* support activities and accomplishments before the day of her decommissioning arrives. Until then, steam on o' mighty *Hudson*.

Photo Captions

Page 1.   *Hudson* steaming through stormy waters during the South Atlantic leg.

Page 2.   Sea foam trail of *Hudson* during a between-station transect.

Page 5.   *Hudson* moving cautiously through the Arctic ice-infested waters.

Page 6.   Decal image of the *Hudson 70* 11 month long voyage that was used for presentation vests given to participants at the 2009 celebration.

Page 7.   (upper): Aerial view of *Hudson* bumping its way through Arctic ice flows.

Page 7.   (lower): Stern view of *Hudson* in 1969 showing current meter mooring floats that were used by oceanographers in the Drake Passage ocean currents study.

Page 8.   Some of the members of the *Hudson 70* Planning Committee.

Page 10.   Rack of Nansen water sampling bottles on the wall of *Hudson's* starboard hydrographic winch room.

Page 11.   (upper left): Retrieval of a Niskin bottle water sampler. (lower left): Deployment of Knudsen bottle showing the three reversing thermometer cases mounted on the bottle and the end caps in the open position. (lower left): Recovery of an acoustic pinger which was attached to the lowering wire just above the lead anchor weight.

Page 12.   Deploying a current meter mooring in Drake Passage. The launching procedure is explained in the text on this page.

Page 13.   (upper left): Deploying an expendable bathythermograph or XBT. (middle right): Retrieving an acoustic release module that was used to release a submerged current meter mooring from its anchor so that it could be recovered. (lower left): Deploying a small air gun. It was used to collect reflection seismic records of seabed sediment deposits.

Page 14.   (upper left): Recovering a Dietz LaFond sediment sampler. (lower left): A suite of rock samples collected from the slope of a volcano on Deception Island. The size and shapes of the rocks are comparable to what might be seen in a typical rock dredge sample. (middle left): Retrieving the VanVeen sediment sampler. (lower left): Night retrieval of the rock dredge sampler.

Page 15.   (left): Boatswain Joe Avery and crew members standing over two core liner sections from a recently-collected piston core sediment sample. (right): A 6 metre long Alpine gravity corer being hauled aboard *Hudson*.

Page 16.   (left): Core cutter end of recently collected piston core showing the bottom surface of the sediment core. (right): Underwater camera module being lowered into the ocean for its journey to the seafloor.

Page 27. (upper left): Geophysicist Jim Shearer and associate monitor reflection seismic and side-scan sonar chart recorders. (upper right): Dr. Peter Beamish tending to his whale sound recorders in *Hudson's* General Purpose (GP) laboratory. (lower right): Fine sand-sized heavy mineral and quartz particles separated from a sediment sample. (middle left): Marine geologists Bernard Pelletier, Gustavs Vilks and an unidentifiable expedition member examine recently collected mud sample. (middle right): Mineral nodules washed from a seafloor sediment sample.

Page 29. (upper): *Hudson's* lifeboat is launched for an evening mission to a near shore shallow water area. (middle): Lifeboat picks its way through sea ice during an Arctic leg deployment. (lower): *Hudson's* lifeboat and survey launch investigate shallow near shore environments near Antarctica.

Page 30: Deception Island reconnaissance team poses for a photo shoot. A young Peter Wadhams is standing on the far left.

Page 31. (upper left): Dr. Iver Duedall operates the hydrographic winch during the daily coffee break. (upper right): Mooring technician Tom Foote adjusts current meter mooring float hardware just prior to its deployment in Drake Passage. (lower left): Crew member and assistant pay out the steel wire rope from the drum of the hydrographic winch. (middle right): Geological technician Mike Gorveatte empties a VanVeen sediment sampler that has just been retrieved. (upper right): Night deployment of an underwater 35 mm camera system. (lower left): Current meter mooring being retrieved from a location in Drake Passage.

Page 32. (upper right): Large mud sample that was collected with the epibenthic sled sea floor sediment sampler. (middle right): Starboard lifting gallows in use during a piston corer recovery operation. The trigger weight release mechanism of the corer can be seen hanging on the wire cable. (upper left): Underwater T.V. and 35 mm camera system used by marine geologists during the Arctic leg of the expedition. (lower left): Defense Research Establishment engineers recover deep scattering layer acoustic transponder. (lower right): Roger Smith washes down a plankton net to recover specimens that have not been transported to the collection canister at the bottom of the net.

Page 33. (left): Scientific staff and ship's officers gather in *Hudson's* Officer's Lounge in between work shifts. (upper right): The small wet bar in the Officer's Lounge offered a convenient place to swap information about *Hudson's* daily scientific and other of the ship's activities. (middle right): During the midnight to 8 AM shift, the Officer's Lounge was usually deserted leaving lots of resting spaces for the ship's doctor's dog. (middle right): Doctor's dog and one sunbather have the upper deck all to themselves, (lower right): The upper deck of *Hudson* was a favorite gathering place during the early evening hours of the warmer months.

References cited and further reading

Edmonds, A., 1973. *Voyage to the Edge of the World.* McClelland and Stewart Ltd., Toronto, 254 p. (ISBN # 0771030673).

Fowler, G., 2002. *Hudson 70* Revisited. *Bedford Institute of Oceanography* 2002 *in Review*, p. 47-48.

Pelletier, B.R., 1988. Marine Science Atlas of the Beaufort Sea: Geology and Geophysics. Misc. Rep.40, Geological Survey of Canada, 41 p.

Smith, R., 2002. *CCGS Hudson* — A Snapshot of Historic Firsts. *Bedford Institute of Oceanography* 2002 *in Review*, p. 44-47.

Schafer, C., Currie, C. and Frobel, D., 2009. Celebrating the 40th Anniversary of the *Hudson 70* Expedition. *Bedford Institute of Oceanography* 2009 *in Review*, p. 46-48.

Wadhams, P., 2009. *The Great Ocean of Truth: Memories of [Hudson 70]*, the first circumnavigation of the Americas. Melrose Press Ltd., St. Thomas Place, Ely, Cambridgeshire, U.K., 378 p.

Wadhams, P., 2009. *Hudson 70*: First Circumnavigation of the Americas. Oceanography, Vol. 22, No.3, p. 227-235.

Weaver, C. and Shedlock, K., 1996. Estimates of seismic source regions from the earthquake distributions and regional tectonics in the Pacific northwest. US Geological Survey Professional Paper 1560, vol. 1, p. 285-306.

Hudson Participants

The whole complement:
scientists, officers and crew

THE CREW

Ahumada, R. (21)

Atkinson, L.P. (9)

Barrett, D.L. (1)

Bary, B. Mck. (10)

Beamish, P.C. (4)

Belanger, J.R. (1)

Belleguay, R. (5)

Bhattacharyya, P.J. (9)

Bertrand, W. (6)

Bluy, O. (7)

Bousfield, E.L. (3)

Brewers, P.G. (16)

Brown, R.G.B. (8)

Bruce, J. (1)

Burt, W.V. (18)

Cameron, W.M. (26)

Carson, B.D. (1)

Choi, C.I. (9)

Chase, K.L. (10)

Chuecas, L.A. (21)

Coady, V. (1)

Conover, R.J. (4)

Cooke, F. (11)

Cook, R.C. (9)

Coote, A.R. (1)

Corbett, T.J. (1)

Corkum, P.L. (1)

Courtney, T.F. (1)

Davidson, R. (10)

Davis, E. (19)

Dean, J.P. (16)

Deevey, G.B. (9)

Densmore, C.D. (16)

Druhan, D. (1)

Duedall, I.W. (3)

Eaton, R.M. (1)

Edwards, R.L. (14)

Faber, D.J. (2)

Foote, T.R. (1)

Freeman, K.F. (4)

English, D. (10)

Garner, D.M. (1)

Gill, J.G. (7)

Gorveatt, M. (1)

Grant, A.B. (1)

Greifeneder, W.B. (1)

Harding, A. (27)

Havard, C. (6)

Haworth, R.T. (1)

Henderson, H. (1)

Hesseler, R.R. (22)

Hiltz, R.S. (1)

Hughes, M. (1)

Hyndman, R. (9)

Inostroza, H.M. (21)

Johnston, B.L. (1)

Keen, C.E. (1)

Lalli, C.M. (13)

Landry, L.P. (5)

Lewis, E.A. (1)

Lister, C.R.B. (19)

Loncarevic, B.D. (1)

MacDonald, R. (10)

Maclean, A. (9)

MacPherson, H.A. (7)

Manchester, K.S. (1)

Mann, C.R. (1)

Markham, J.W. (2)

Maunsell, C.D. (1)

McHughen, S.B. (1)

Melanson, R.C. (1)

Michael, A.D. (9)

Mills, E. L. (9)

Montaner, R.E. (22)

Muise, F. (1)

Murray, J.G. (1)

Neilsen, J.A. (1)

Paranjape, M.A. (4)

Parker, R. (6)

Pelletier, B.R. (1)

Pickard, G.L. (10)

Piechura, J. (1)

Pilote, J.M.R. (1)

Pinner, W. (15)

Pond, G.S. (18)

Prakash, A. (4)

Probert, P. (20)

Purdy, G.M. (1)

Rankin, D. (9)

Rebaudi, R.S. (25)

Reiniger, R.F. (1)

Rey, F.R. (21)

Rock, N.R. (24)

Rogers, G. (26)

Ross, D.I. (1)

Sacks, P.L. (16)

Schafer, C.T. (1)

Sharpe, J.H. (9)

Shearer, J. M. (1)

Sheldon, R.W. (4)

Shih, K.G. (1)

Silva, N. (23)

Smith, R. (11)

Solowan, P.A. (1)

Srivastava, S.P. (1)

Steele, J.P. (1)

Storm, M.P. (10)

Strad, J. (15)

Sutcliffe, W.H. (4)

Taylor, B. (1)

Thomlinson, A. (1)

Tiffin, D. (5)

Toom, H. (1)

Uccelletti, B.D. (22)

Vilks, G. (1)

Von Arx, W. (16)

Wadhams, P. (1)

Wagner, F.J.E. (1)

Wangersky, P.J. (9)

Watt, W. (9)

Whiteway, W.J. (1)

Winters, D. (1)

Wong, C.S. (5)

Woodside, J.M.

Yorath, C. (6)

Yoshinari, T. (9)

Zurbrigg, R.E. (12)

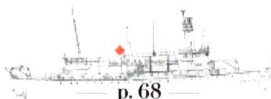

AFFILIATIONS

(1) Atlantic Oceanographic Lab., Bedford Institute, Dartmouth, NS.

(2) National Research Council of Canada, Ottawa.

(3) National Museum of Natural Sciences, Ottawa.

(4) Marine Ecology Lab., Bedford Institute, Dartmouth, NS.

(5) Pacific Oceanographic Group, Nanaimo, BC.

(6) Geological Survey of Canada, Ottawa.

(7) Defense Research Establishment Atlantic, Dartmouth, NS.

(8) Canadian Wildlife Service, Bedford Institute, Dartmouth, NS.

(9) Institute of Oceanography, Dalhousie University, Halifax, NS.

(10) Institute of Oceanography, University of BC, Vancouver, BC.

(11) Queen's University Kingston. Ontario.

(12) University of Toronto, Toronto, Ontario.

(13) McGill University, Montreal, Quebec.

(14) Trent University, Peterborough, Ontario

(15) Highland Helicopter Ltd., Vancouver, BC.

(16) Woods Hole Oceanographic Institution, Woods Hole, MA., USA.

(17) Scripps Institution of Oceanography, La Jolla, California, USA.

(18) Oregon State University, Corvallis, Oregon, USA.

(19) University of Washington, Seattle, Washington, USA.

(20) Royal Navy

(21) Universidad de Conception, Chile.

(22) Instituto Hidrografico, Chile.

(23) Universidad Catiolica, Valparaiso, Chile.

(24) Instituto Oceanografica Sao Paulo, Sao Paulo, Chile.

(25) Argentine Navy, Buenos Aires, Argentina.

(26) Marine Sciences Branch, Energy, Mines & Resources, Ottawa.

(27) Hunting Ltd., United Kingdom.